The Facts on Matter

What are THE STATES OF MATTER?

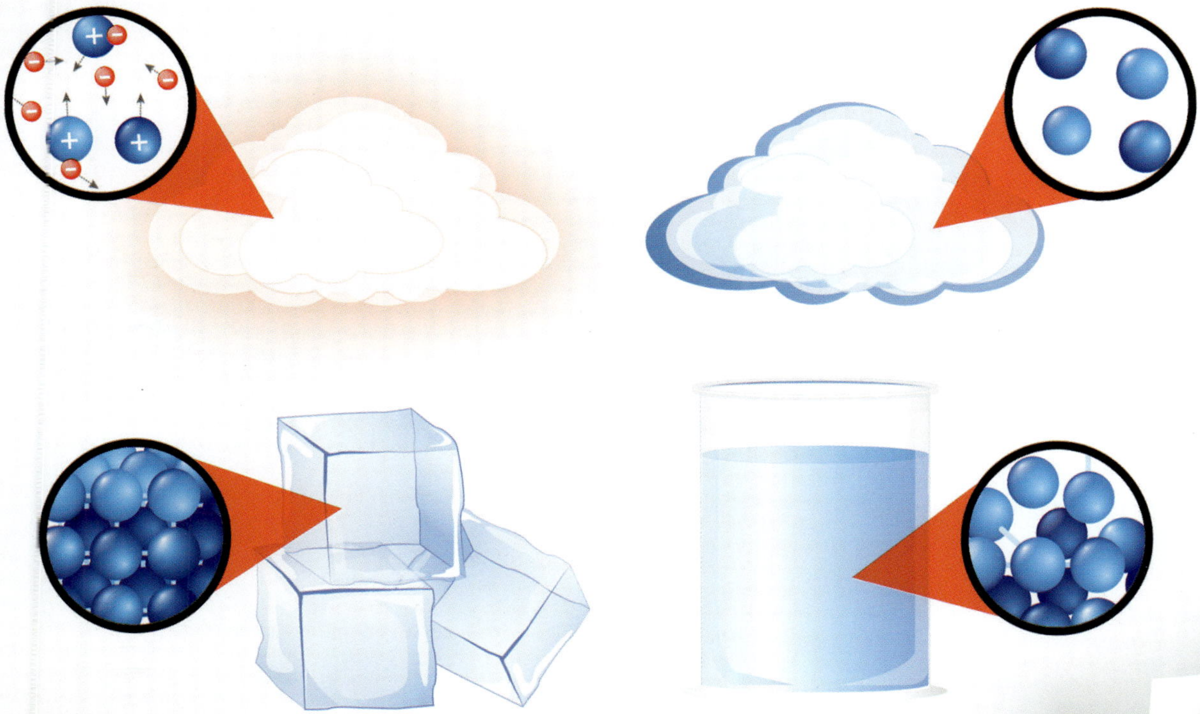

By Nathan Miloszewski

Gareth Stevens PUBLISHING

Please visit our website, www.garethstevens.com. For a free color catalog of all our high-quality books, call toll free 1-800-542-2595 or fax 1-877-542-2596.

Library of Congress Cataloging-in-Publication Data

Names: Miloszewski, Nathan, author.
Title: What are the states of matter? / Nathan Miloszewski.
Description: New York : Gareth Stevens Publishing, [2022] | Series: The facts on matter | Includes index.
Identifiers: LCCN 2020042796 (print) | LCCN 2020042797 (ebook) | ISBN 9781538267219 (library binding) | ISBN 9781538267196 (paperback) | ISBN 9781538267202 (set) | ISBN 9781538267226 (ebook)
Subjects: LCSH: Matter–Properties–Juvenile literature.
Classification: LCC QC173.36 .M58 2022 (print) | LCC QC173.36 (ebook) | DDC 530.4–dc23
LC record available at https://lccn.loc.gov/2020042796
LC ebook record available at https://lccn.loc.gov/2020042797

First Edition

Published in 2022 by
Gareth Stevens Publishing
29 E. 21st Street
New York, NY 10010

Designer: Katelyn E. Reynolds
Editor: Char Light

Photo credits: Cover, p. 1 Designua/Shutterstock.com; cover, pp. 1, 17 (diagram) ttsz / iStock / Getty Images Plus; cover, pp. 1–24 (background) sudanmas/E+/Getty Images; cover, pp. 1–24 (chemistry doodles) backUp/Shutterstock.com; cover, pp. 1–24 (banner) Ozz Design/Shutterstock.com; cover, pp. 1–24 (paper) Sergey Mironov /Shutterstock.com; pp. 4, 7, 9, 11, 13 (diagram) VectorMine/ iStock / Getty Images Plus; p. 5 (diagram) blueringmedia/ iStock / Getty Images Plus; p. 5 XiXinXing/ Getty Images; p. 7 Taiyou Nomachi/ DigitalVision/Getty Images; p. 9 JGI/Jamie Grill/Getty Images; p. 11 Peter Correz/ The Image Bank / Getty Images Plus; p. 13 Ronald Verhage/Shutterstock.com; p. 15 robuart/ iStock / Getty Images Plus; p. 17 Kathy Collins/ Photographer's Choice / Getty Images Plus; p. 19 elenabs/ iStock / Getty Images Plus; p. 21 Kira Kreutzbender / EyeEm/Getty Images.

Printed in the United States of America

Some of the images in this book illustrate individuals who are models. The depictions do not imply actual situations or events.

CPSIA compliance information: Batch #CWGS22: For further information contact Gareth Stevens, New York, New York at 1-800-542-2595.

CONTENTS

Words in the glossary appear in **bold** type the first time they are used in the text.

What Is MATTER?

Everything you can see around you is made up of matter, from the shoes on your feet to the trees in the park. Even the wind is made of matter! Matter is made up of atoms. Atoms are the smallest, most basic parts of a **substance** that can exist on their own.

Matter is found on Earth in three main states: solid, liquid, or gas. There is a fourth state of matter called plasma. Read on to find out more about each state!

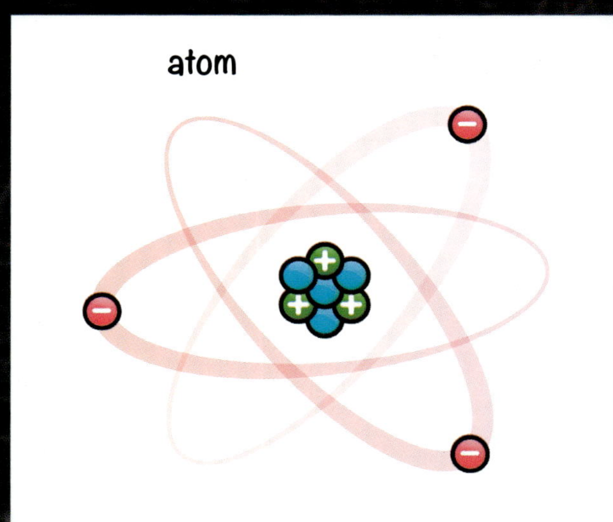

At lunchtime, your drink is always going to be a liquid. Your sandwich will always be a solid!

What Is a SOLID?

Solids are easy to recognize because they're hard to the touch. They keep their shape no matter what kind of container they're in. Trees, pieces of fruit, and rocks are all solids.

In solids, atoms are packed in tight together. They don't move around a lot. This helps them keep their shape. A solid's shape can change when **force** is used. To change the shape of a solid, you need to break it, cut it, or melt it with heat. Some solids are harder to change the shape of than others!

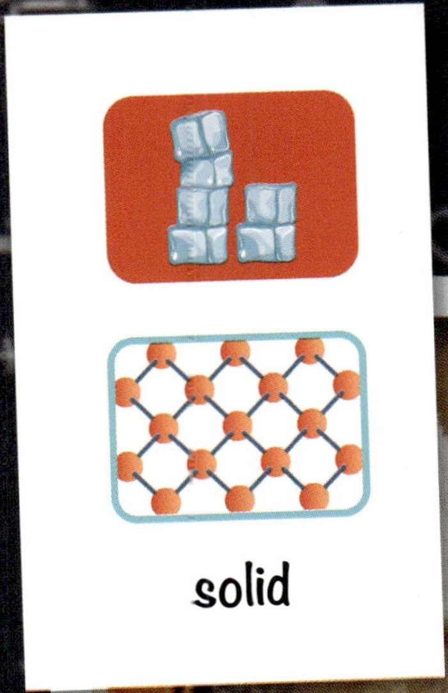

To change the shape of a tree, you need to use force and a tool like a saw to cut it down into smaller pieces.

What Is a LIQUID?

A liquid is the type of matter that can get you wet, like water. In fact, most of the liquid on Earth is water! Liquids flow easily. If you pour one into a cup or a bucket, the liquid takes the shape of that container.

The atoms in a liquid are fairly close together, and they stay near one another as the liquid goes from one place to another. But the atoms in a liquid move around more than those in solids. This lets the liquid flow and change shape.

Know the Basics!

Mercury is the only metal in liquid form at room temperature.

liquid

Lemonade is a liquid. When you pour it in a glass, the lemonade takes the shape of the glass.

What Is a GAS?

Most gases are invisible or can't be seen by the human eye. Gases don't have a shape, and they are like liquids because they spread out to fill whatever container they're in.

The atoms in a gas aren't close together and move around easily. If you don't trap a gas, it won't stop spreading! Putting the gas helium in a balloon will make the balloon float. But if you let the helium into your backyard, it would spread out and, to our eyes, disappear.

Know the Basics!

People breathe in oxygen and breathe out carbon dioxide. These are both gases!

gas

Air is a mixture of gases. You may not be able to see air but without it, you wouldn't be able to glide through the sky with a parachute—or breathe!

What Is PLASMA?

The fourth state of matter is the most common throughout the **universe**: plasma. Plasma starts as a gas. When enough **energy** is added to the atoms of a gas, it becomes plasma. Plasma is very hot!

You **interact** with liquids, solids, and gases all day long. But you likely see plasma around you too! Some TVs use plasma to show you your favorite show. Plasma is also found in the lights of neon signs.

plasma

The lightning you see on a stormy night is a type of plasma!

The Study of MATTER

The study of matter is part of the **physical** sciences. Chemistry studies why matter is made the way it is and how matter can change. You're actually performing chemistry when you bake a cake or cook dinner! You're mixing different foods together and seeing how they interact.

Physics is the study of force and matter in motion. Sir Isaac Newton is a famous physicist who explained how the force of **gravity** works on objects. Astronomy studies everything in the universe using physics and chemistry to explain how the stars and planets move around in space.

Different fields of science study the ways different kinds of matter **behave** and interact with other matter.

Transforming States of MATTER

Matter can change into a different state when energy is added or taken away from the atoms in that matter. When energy is added to a solid, it can change to a liquid. This energy is often in the form of heat. Liquid that is heated can turn into a gas. This is called evaporation.

When energy, like heat, is taken away from atoms, liquid matter can become a solid. An example of this is freezing water to make ice cubes. Condensation is the process in which a gas, such as **water vapor**, loses energy and changes back to a liquid.

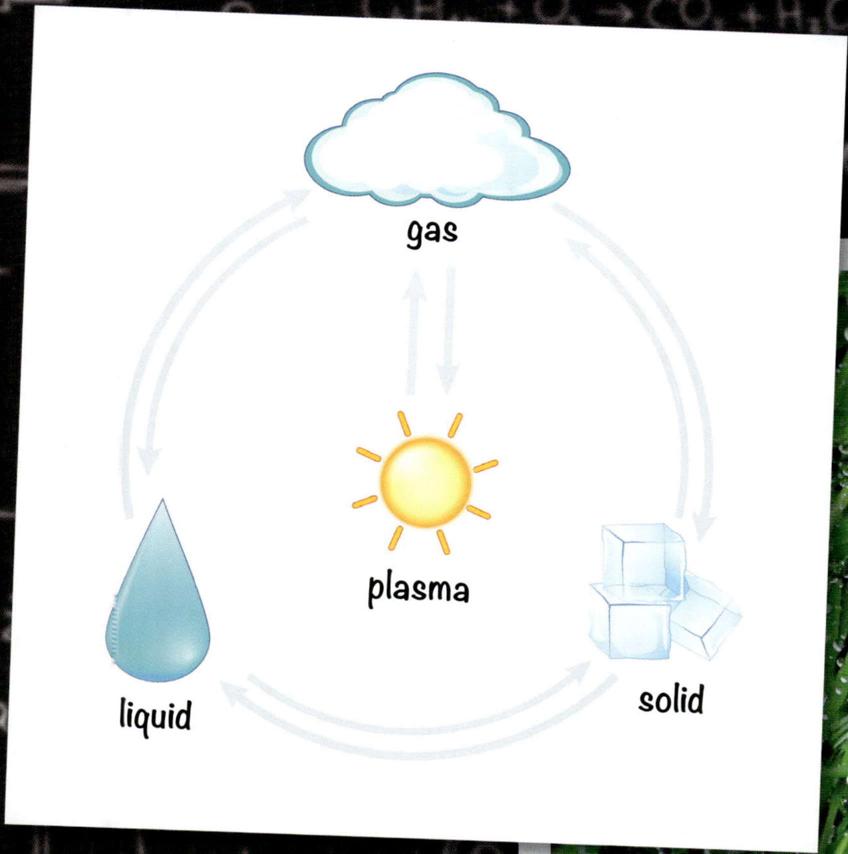

On a cool morning you may see dew on the grass. Dew is water from condensation, which occurs when water vapor cools overnight.

some of it evaporates and rises into the air. Condensation occurs and forms clouds. When the water in clouds warms up, it becomes liquid again and falls to Earth as **precipitation**.

Know the Basics!

Runoff is the water that flows over Earth after it falls as precipitation. It may end up in streams or lakes, or soak into the soil.

THE WATER CYCLE

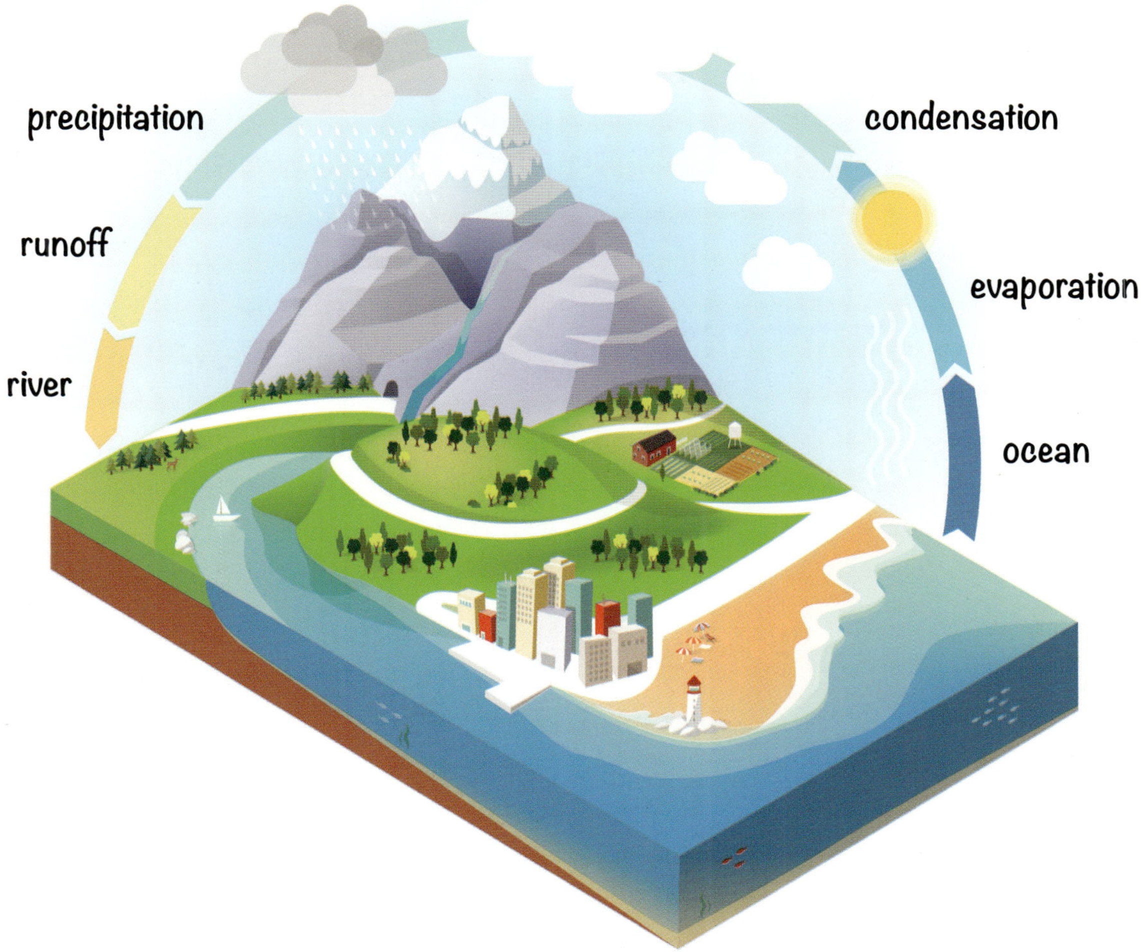

The water cycle is always happening!

Making S'MORES

Matter may go through physical changes and chemical changes. Physical changes include changes to size, shape, or the state of matter. They don't change the atoms of the matter. A chemical change is when the matter becomes a new kind of matter.

When you're making s'mores and the chocolate melts, that's a physical change. When you burn the marshmallow and see parts of it have turned black, the fire has caused a chemical change!

Know the Basics!

Burning the marshmallow causes its **molecules** to change into something new—the element carbon!

Melted chocolate is still chocolate—that's just a physical change!

GLOSSARY

behave: to act a certain way

energy: power used to do work

force: power or effect. Also, a natural power or effect that can change the speed or direction of something.

gravity: the force that pulls objects toward Earth's center

interact: work or act together

model: a picture or set of steps used to explain a scientific idea

molecule: two or more atoms that are joined, or bonded

physical: having to do with the natural world

precipitation: water that falls onto the ground such as rain

substance: a certain kind of matter

universe: everything that exists

water vapor: very small drops of water mixed with air in the form of a gas

For More INFORMATION

Books

Biskup, Agnieszka. *The Solid Truth about States of Matter with Max Axiom, Super Scientist: 4D An Augmented Reading Science Experience*. North Mankato, MN: Capstone Press, 2019.

Diehn, Andi. *Matter: Physical Science for Kids*. White River Junction, VT: Nomad Press, 2018.

Troupe, Thomas Kingsley. *What's the Matter with the Three Little Pigs?: The Fairy-Tale Physics of Matter*. North Mankato, MN: Picture Window Books, 2018.

Websites

PBS Heating and Cooling
ny.pbslearningmedia.org/resource/solid-or-liquid-media-gallery/hero-elementary/
A student accidentally freezes a lake—see how a team of friends get it back to its liquid state.

States of Matter
abcya.com/games/states_of_matter
An educational activity that introduces the different states of matter and helps you learn about their properties.

States of Matter Science Experiments
stevespanglerscience.com/lab/categories/experiments/states-of-matter/
Learn how to use things around the house to create your own states of matter experiments.

Publisher's note to educators and parents: Our editors have carefully reviewed these websites to ensure that they are suitable for students. Many websites change frequently, however, and we cannot guarantee that a site's future contents will continue to meet our high standards of quality and educational value. Be advised that students should be closely supervised whenever they access the internet.

INDEX